Heinkel He 112

Hans-Peter Dabrowski

He 112 B-0 export machines requisitioned by the Reichsluftfahrtministerium(RLM) during the Sudeten crisis. The wing undersides show a part of the typical Heinkel factory markings (HE+..)

Schiffer Military/Aviation History
Atglen, PA

Selected Sources:

Ernst Heinkel: "Stürmisches Leben", Mundus-Verlag, Stuttgart 1953

H. Dieter Köhler: "Ernst Heinkel - Pionier der Schnellflugzeuge", Bernard & Graefe Verlag, Coblenz 1983

Bruno Lange: "Typenhandbuch der Deutschen Luftfahrttechnik" Bernard & Graefe Verlag, Coblenz 1986

William Green: "Warplanes of the Third Reich", Doubleday and Company Inc., Garden City, New York 1973

Karl Ries: "Luftwaffenstory 1935-1939" Verlag Dieter Hoffmann, Mainz 1974

Fred Haubner: "Die Flugzeuge der Österreichischen Luftstreitkräfte vor 1938", Weishaupt Verlag, Graz

Friedrich Ritz: "Mein Leben" (unpublished)

"Der Jäger an der Kette" von Gert W. Heumann (Flugrevue 5-6/1960)

"The Curious Saga of the He 112" - Air International 5-6/1989

"IV Internationales Flugmeeting Zürich" - Der Deutsche Sportflieger 7/1937

"He 112 B-1" Heinkel leaflet, Stuttgart

"Die Heinkel He 112 in spanischen Diensten" by Harold Thiele and Juan Arráez Cerdá - Jet & Prop 2/94

Heinkel "Exposé für Besprechung über Export-Fragen" 8/18/1938

RLM LC 6/38 Ref. Ministerialrat Mueller, response to memorandum, 8/27/1938, classified

Internal Heinkel memorandums, briefs, travel reports, sources from 1933-1937

Prof. Dr.-Ing. Heinrich Hertel: Erinnerungen an die Arbeiten zur Anwendung von Flüssigkeitsraketen im Flugzeugbau und an die Entwicklungen mit Wernher von Braun, DGLR reports 1/2 1970, März

MG C/30-I, Waffen Revue 55 IV/84

PHOTOS:

Bernád, Cerdá via Thiele, Conrad, Griehl, Hartwig, Heinkel Archives, Koos, Lange, Matthiesen, Meier, Nowarra, Petrick, PRO, Ritz, Roosenboom, Schnittke, Selinger

ACKNOWLEDGMENTS

I would like to express my heartfelt thanks to all those who, in both word and deed, have contributed to this compilation, particularly Messrs. Siegfried Fricke, Dr. Volker Koos, Hans Justus Meier, Helmuth Roosenboom, Hanfried Schliephake and Harold Thiele, plus Ralf Modemann for his technical assistance.

A REQUEST

In researching the He 112 I've noticed several contradictions, and this compilation leaves a series of questions unanswered. The authentic documents from Heinkel's archives show, where available, the history of the prototypes to be somewhat different than that which can be found in popular aviation literature. Would any reader who can provide additional documented information regarding the He 112 or is aware of errors in this book please contact me personally or via the publisher.

Hans-Peter Dabrowski

Three He 112 B-1s, here without civilian markings and only wearing the Balkan cross. It is unknown what role these aircraft played.

Translated from the German by Don Cox

Printed in China.
ISBN: 0-7643-0392-9

This book was originally published under the title, *Waffen Arsenal-Jagdeinsitzer He 112: Konkurrenz zur Me 109 und Exportflugzeug* by Podzun-Pallas Verlag.

We are interested in hearing from authors with book ideas on related topics.

Published by Schiffer Publishing Ltd.
4880 Lower Valley Road
Atglen, PA 19310
Phone: (610) 593-1777
FAX: (610) 593-2002
E-mail: Schifferbk@aol.com
Please write for a free catalog.
This book may be purchased from the publisher.
Please include $3.95 postage.
Try your bookstore first.

INTRODUCTION

A NEW STANDARD FIGHTER

With the exception of a handful of the more famous types, the general public is virtually unaware of those German aircraft types produced up until 1945. These familiar types include the "Tante" Ju 52 or the Me 109 (of which there are still a few examples flying today), the He 111 and the Ju 88. Or maybe a person has heard or read of the legendary "secret weapons", which at that time were kept under lock and key (as were all the others) and confiscated by the victors - who also kept them under wraps for a time.

There was a considerable number of various aircraft types in the period from about 1923 to 1945, naturally including "exotic" and adventurous designs without a future, other designs dropped due to lack of funds, prototypes, etc. On the one hand, from just 1933 to 1945 serious developmental work was undertaken which for whatever reasons was not recognized or acknowledged. On the other hand, those who had the power within the Third Reich passed off technical achievements as the norm; record flights were usually made by "production aircraft".

Ignorance and arrogance went hand in hand in the power control centers of the day. Thus, they often supported or prevented/blocked individual developments, disregarding actual requirement estimations. One of those designs which was prevented/blocked was the He 112 fighter, although at least part of the blame for its lack of success rested with the technical management at the Heinkel Flugzeugwerke at the time.

At the end of 1933, the Army's requirement read:

Development of a single seat fighter for day and night air operations, single engine, 400 km/h at 6,000 m altitude, armament either two machine guns with 1,000 rounds per barrel or one 20 mm machine cannon with 200 rounds. In addition, the Army laid out other detailed requirements, although these won't be explored within the pages of this book.

Heinkel submitted the Project P 1015, a rugged low wing design by Siegfried Günter. The wings had double spars, so that even if damaged a load multiple of four could be guaranteed. The aircraft was to have resembled the He 70 in appearance, meaning with curved elliptical wings, a feature which required a considerable amount of work time.

Although the design initially called for an enclosed pilot's cockpit, the final version had an open cockpit (although a sliding canopy could be retro-fitted without difficulty). In addition to water cooling, evaporative cooling was also to have been utilized.

Fuselage model of the He 112 in the wind tunnel.

The first "Me 109", D-IABI, at the time still a Bf 109 of the Bayerische Flugzeugwerke (BFW). The aircraft, designed by Willy Messerschmitt, won the standard fighter competition.

Right: Fellow competitor and loser: Arado Ar 80 (here the V2). The landing gear did not retract - unacceptable for a modern fighter aircraft in the new Luftwaffe.

Also without a chance: The Fw 159 was developed from the Fw 56 Stößer. As a high wing design with complicated retractable undercarriage, the design was considered to be unsuitable for further development and was duly rejected.

The He 112s wings seen from below.

Left: All-metal fuselage of the He 112.

Fuselage with the engine already fitted and a complete wing assembly, being mated together. This is either a prototype or an He 112 A-0, due to its lack of canopy.

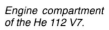

Engine compartment of the He 112 V7.

Right: Production of the He 112 for the export market.

Fuselage stress testing on an He 112.

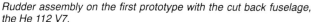
Rudder assembly on the first prototype with the cut back fuselage, the He 112 V7.

View into the cockpit of one of the He 112 V-machines.

A Heinkel memorandum dated 21 July 1934 instructed that studies be made of the Rolls Royce Kestrel, BMW 115 and 116 and the Jumo 10 engines with regard to evaporative cooling. This type of cooling was subsequently utilized on the He 100 V8 (disguised under the designation of He 112 U) when it set the world's absolute speed record.

The aircraft's rugged construction was apparent at first glance; it possessed perfectly acceptable flight handling characteristics and easily met all the requirements. But it had competition: in addition to the Arado Ar 80 and the Focke-Wulf Fw 159 there was also the Bf 109 from the Bayerische Flugzeugwerke (designed by Willy Messerschmitt, Heinkel's strongest rival in the battle for a large-scale production contract for the future standard fighter). The Ar 80 proved to be too heavy for the light design requirement, plus it had a fixed spatted landing gear. It was quickly ruled out, as was the Fw 159. The Fw 159, with its unique and rather complicated retractable undercarriage, was classified as not being suitable for further development. The Messerschmitt Bf 109 was faster, lighter, more simple of a design and cheaper to produce than the He 112. Accordingly, production was also cheaper, faster and easier, as were maintenance and repairs. Later, in May of 1937, Heinkel also came to the conclusion that the He 112 required nearly twice as much time to build than Messerschmitt's airplane. This type had an enclosed cockpit and retractable landing gear. The Rechlin test center criticized the Bf 109 in part for its weak undercarriage - prone to causing ground loops - and the side opening canopy.

Although the He 112 may have been able to accumulate more plus points, numerous changes caused the prototype to lag further and further behind the Messerschmitt Bf 109.

The older and experienced pilots Udet and Ritter von Greim, plus two young Travemünde test center pilots, Francke and Conrad, flew the first prototypes of the competing firms in October of 1935 in Travemünde. After a series of detailed evaluations they all decided in favor of the Messerschmitt Bf 109. Powered by a Jumo 210 engine, Hermann Wurster flew the Bf 109 V2 during the final comparison phase on 15 April 1936. He put the aircraft into a dive from 7,000 meters, pulling up just above the ground and subjecting the aircraft (and himself) to 7.8 g. He easily spun the airplane with 21 right and 17 left rotations. Gerhard Nitschke flew the He 112 V 2 (also driven by a Jumo 210). The aircraft was lost in a crash during the spin demonstration.

In his accident report, Nitschke wrote the following:

"After I'd completed the dive and right spins, I climbed back up to altitude and began putting the aircraft into left spins at around 2,400 meters. The aircraft went into a smooth stall and dropped over into a flat spin." The experienced pilot tried every trick in the book to get the airplane out of the spin, but when the engine died and then froze, Nitschke was forced to bail out at around 500 meters. Although the tumbling aircraft damaged the primary parachute's auxiliary 'chute, Nitschke landed safe and sound on the southeast corner of the airfield. The machine impacted near the levee; no one was hurt.

This photo shows the accessibility of one of the MG 17s installed in the He 112, with all appropriate access panels opened.

The aircraft later known as the Me 109 was therefore declared the winner as a result of the series of evaluation flights; the die had been cast. Heinkel's He 112 eventually was given the status of a backup design, although it was determined that no second standard fighter was to supplement the Me 109 then entering production. When problems arose with the further development of the Me 109, Heinkel was consequently not in an immediate position to exploit the situation and prove to the RLM that an error in judgement had been made—at least in the fixation with just a single type (a view which changed during the war with the introduction of the Fw 190).

Ernst Heinkel pinned the blame for his defeat on his developmental chief, Dr. Heinrich Hertel, who just before the outbreak of the Second World War transferred to Junkers. In Heinkel's biography "Ein stürmisches Leben" he wrote: "...But Hertel was the one who, when building the He 112, from the start brought along his tendency for endless changes, experimentation and novelties and in doing so, was the final reason that the He 112 was really completed after the Me 109..."

In order to remain competitive in the export market and offer an aircraft on par with the Me 109, the He 112 had to be made lighter. A new, shorter wing was finished in the spring of 1936, but did not result in any appreciable improvement in performance. Additional V-prototypes continued up to V6 and V8, and there followed a small A-0 series of just four aircraft. With the V7, which was virtually a new design (new fuselage and wings, new undercarriage, new empennage, enclosed cockpit, a shorter, three-piece bubble canopy), the He 112 had virtually reached parity with the Me 109 (which in the interim had also seen improvements). By this time, however, Messerschmitt's design was already being produced in series.

Heinkel test pilot Friedrich Ritz after a failed attempt at clearing the soot away from the exhaust path: due to a broken pipe the soot was dumped directly into the cockpit instead of along the sidewall (probably He 112 V4).

The first He 112, powered by a Rolls Royce Kestrel S II engine and lacking the retractable radiator. The only armament which could be carried by the V1 was six 10 kg fragmentation bombs; machine guns had not yet been incorporated into the design.

THE VARIOUS PROTOTYPES OF THE HE 112

Eleven prototypes, a small A-series and later B-0 and B-1 series of the He 112 were produced; export variants were given the designation letter E. The A and B designations originally stood for open and enclosed cockpit versions, respectively; only later did they reflect series production variants. As early as October of 1933 Heinkel had a Project P 1015 listed for a V.J. (Verfolgungsjäger, or interceptor, an airplane which could pursue and engage enemy bombers or reconnaissance aircraft - the common designation for fighters at the time), armed with an MG C/30 L (2 cm). On 2 December 1933 the decision was made to fit this aircraft with a BMW XV engine or "if necessary, with a Rolls Royce Kestrel engine". The design description, dated 2 May 1934, indicated that the project was classed as a "single seat sportplane", On 5 May 1934 P 1015 became He 112 and on 8 June 1934 the mockup was displayed for the first time.

The He 112 was built with two different fuselage designs: round taper all the way back to the tail tip (V1 through V6 and V8, plus the A-series), and running in a so-called "vertical cut" back to the tail end (P 1015, V7, V9 through V11 and all other production variants). All versions up to V4 were designed to be armed with MG C 30/L, MG 17 and MG FF in various configurations. In addition, six 10 kg fragmentation bombs could be carried in what were called "Elvemags" (elektrisch ausgelöstes Vertikalmagazin, or electrically activated vertical rack). As far as is known, the following information is provided for the prototypes and those aircraft in the A-series:

V1, Werknr. 1290, registration D-IADO, fitted with a British Rolls Royce Kestrel II S engine (700 hp), HKW two-blade propeller (initially fixed, later variable pitch when stationary). Built in the summer of 1935, maiden flight by factory test pilot Gerhard Nitschke on 1 September 1935. Factory test flown in Rostock. During evaluation flights on 12 and 13 November 1935 in Marienehe by Travemünde test facility pilots Blume, Thoenes and Trenkmann the several points came under criticism: e.g. windscreen useless (poor visibility, limited protection), seat too low (visibility!), retracting the landing gear by hand pump requires too much effort (40-50 strokes!), etc. On the other hand, the flight handling characteristics were found to be generally satisfactory. In October of 1935 the He 112 V1 took part in the flyoff in Travemünde. In April the V1 was fitted with new, shorter wings, giving it a wingspan of 11.5 meters. The old wings were later used for stress testing. Later the V1 was flown as a calibration platform for barometric pressure calibration. In May of 1940 the machine was registered as TH+HW, its subsequent fate being unknown.

V2, Werknr. 1291, registration D-IHGE, powered by Jumo 210 C (600 hp), HKW three-blade propeller, variable pitch when stationary.

Above and left: He 112 V1 D-IADO prior to painting, during testing of the liquid-cooled Rolls Royce twelve-cylinder V engine.

Below: The V1 seen here in its finished and painted state. The fixed two-blade propeller was supplied by the firm of HKW (Heidenheimer Kupferwerke, Frankfurt am Main).

Factory test flown (initially with the old wings from V1, wing-span 12.6 m) in Rostock, evaluation in Rechlin. Wingspan reduced to 11.5 m based on data from V1. After nosing over with Friedrich Ritz at the controls in early February 1936 it was stripped of paint. Total loss during spin demonstrations in Travemünde on 15 April 1936. The pilot was Gerhard Nitschke, who bailed out in a life-threatening situation. The machine crashed into the levee and, after being recovered, was declared a total write-off.

As the RLM had informed the Heinkel firm that the He 112 (and He 118) was being considered for carrier operations (for the German aircraft carrier Graf Zeppelin) the prototype was simultaneously evaluated for carrier applications—fitting with a hook, auxiliary wing flaps, descent rate monitors, etc. These may have been undertaken from the V3 onward. For its part, the RLM let it be known that there was no call for a dedicated carrier airplane design; current types were to be simply modified for the new role.

V3, Werknr. 1292, registration D-IDMO, also powered by Jumo 210 C, wingspan as V2 at 11.5 m, Schwarz three-blade propeller (adjustable when stationary). This machine originally had an open cockpit, but was later fitted with a sliding canopy. Factory test flown in Rostock, evaluated in Rechlin. Heinrich Beauvais, Rechlin's chief test pilot, noted a flight in the V3 on 9 January 1937. Following that, the aircraft served as a testbed for rocket propulsion evaluations (Wernher von Braun test company, Neuhardenberg/

Brandenburg). It was apparently destroyed while serving in this capacity.

V4, Werknr. 1974, registration D-IDMY (the flight book of factory pilot Kurt Heinrich shows the first flight as 24 June 1936, but as D-IPMY). Engine was a Jumo 210 D. The aircraft was contracted for by the RLM on 23 May 1936 with the caveat: "The recipient of the aircraft will be made known after construction is completed". In the type inventory listing from 21 June 1937 it states: "Delivered to Special Area Künzel" (as was the V3, by the way). Walter Künzel was an aviation expert and liaison engineer for cooperative efforts between the Heinkel firm and the rocket builder Wernher von Braun. Therefore, there is the distinct probability that Heinkel made the V4, like the V3, available for rocket engine installation without taking a roundabout way.

In December of 1936 Heinkel began top secret construction work on the He 176 rocket powered airplane. In the autumn of 1937 the design was ready for its engine. He 112 prototypes were used for testing the Walter rocket engine, which also fits the time period in question. Test pilot Erich Warsitz had carried out all the important flight tests with this type of powerplant, which up to this point had never been successfully tested. According to the chief of development at Heinkel at the time, Dr. Heinrich Hertel, the testing was broken down into the following stages: 1) He 112 with built-in rocket engine turned this secondary engine on at around 2,000 meters altitude with propeller running and providing main power.

He 112 V2 was fitted with an MG C/30 L 20 mm engine cannon firing through the propeller hub.

Crash landing by Friedrich Ritz in early February 1936 while flying He 112 V2 due to fuel tank switching failure (construction flaw). It's easy to see how the soft plowed field prevented a smooth rollout - the aircraft flipped over.

The injured pilot had to be dug out from underneath the overturned machine by farmers who had rushed to the scene. The aircraft itself suffered only superficial damage.

2) Rocket provided main thrust at approx. 2,000 m altitude, then He 112's propeller switched off. 3) Rocket as sole powerplant during takeoff. All tests were said to have been accomplished without serious problems or accidents occurring.

It is highly probable that V5 through V8 were pulled from the A-0 series, since the V4 had been planned as the prototype model for the A-series.

V5, Werknr. 1951, registration D-IIZO, Schwarz three-blade propeller. Aircraft planned for the so-called "high speed" wing (16 m2) and with wing cannons plus two machine guns. Was flown by Wilhelm Leo Conrad on 6 and 7 October 1936 from Travemünde via Rechlin to Berlin-Staaken for unknown reasons. After testing the V5 was cleared by HWaA in March of 1937 for rocket engine testing.

V6, Werknr. 1952, registration D-IQZE, Schwarz three-blade propeller. The RLM expressly called for a "flawless factory finish" for this prototype.

The demise of He 112 V2: test pilot Gerhard Nitschke was forced to abandon the aircraft and let it crash after it went into a flat spin.

Left: It can be seen that the V2 was not given a new paint scheme during its overhaul in February/March of 1936.

The V2 must have impacted the water nose first. The fuselage of the rugged single seat fighter shows little external damage.

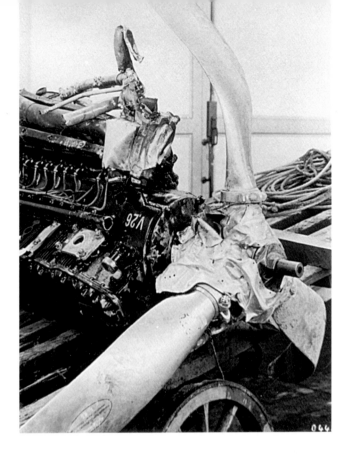

Wings and empennage built using the explosive riveting process. Emergency landing made on 1 August 1936 with pilot Kurt Heinrich at the controls. During takeoff, the rudder became jammed to the right at an altitude of about four meters; company buildings lay in the flight path and the pilot therefore decided to make an immediate landing. During the rollout, the machine ran into the perimeter fencing and was damaged. Fitted with 20 mm MG C/30 L cannons at the Travemünde test center prior to being put through prototype evaluation. Flown by Conrad on 2 October 1936 from Travemünde to Wunstorf and on 5 October 1936 back to Travemünde. Reason unknown, possibly gun testing. Operated in Spain with VJ/88 fighter evaluation squadron of the Condor Legion as 5•1. Flown by pilots Radusch, Balthasar and Schulz. The aircraft was a total write-off after a crash landing on 19 July 1937. The wreckage was transported back to Germany to be examined.

V7, Werknr. 1953, registration D-IKIK, planned to carry three MG 17 guns. Fitted with DB 600 A engine (900 hp), Schwarz propeller, three-blade. First prototype for the B-series with revised fuselage compared to the A-prototypes and fitted with a shortened, glazed fairing over the headrest. Heinrich Beauvais flew the aircraft on 26 June 1937 in Rechlin. The V7 also put in an appearance at the flying meet in Dübendorf, but was not shown in the static viewing area. Exhibiting this aircraft in Switzerland must have been very important for Heinkel (exports!), for Hertel (chief of development) personally applied pressure in written form to all those departments involved in the project. The airplane eventually went on to serve as an engine testbed.

Engine and propeller from the crashed V2. The barrel of the MG C/30 L cannon protrudes from the bent VDM RS propeller. The manufacturing branch of HKW fused with the VDM company in 1931/32, which meant that for a time it was possible find mention of both HKW as well as VDM.

The He 112 V3 D-IDMO was laid out to take three 7.92 mm MG 17s. The aircraft was reported to have been delivered to "Special Area Künzel" (rocket propulsion testing) in March of 1937.

Right: Two views of the modified He 112 V3: the machine was initially flown with an open cockpit, but was later fitted with a small sliding canopy. Apparently, for rocket propulsion trials both the canopy and extended headrest fairing/spine were removed (below). Due to vibration caused by the rocket motor, the machine was fitted with additional bracing which joined the vertical and horizontal stabilizers. The tapered shape of the fuselage tip was certainly well suited for installation of the new powerplant.

The bare He 112 V4. According to Heinkel's records from 1937 it was given the registration code of D-IDMY; in the flight book of test pilot Kurt Heinrich, however, it is noted as D-IPMY. Werknr. 1974 is the same for both. Who has made the mistake here?

Right: This photo of the V4 served quite well as the basis for the "art impression" of an He 112 rocket powered aircraft (compare page 26) and is quite often published as the real thing. The V4 actually went to Special Area Künzel for special equipment installation (rocket engines) and was the only prototype to carry no armament or bomb racks.

Little is known about V5 D-IIZO. It was relatively lightly armed with two MG 17s firing through the propeller arc, apparently being intended for sale to Japan.

He 112 V5 seen from the rear. The aircraft is anchored to the ground with wing tie-downs.

Left: He 112 V6 in Tablada, Spain, for combat evaluation in December 1936. It flew ground attack missions with the coding 5•1 while operating with the Condor Legion.

On 19 July 1937 the V6 "Kanonenvogel" (Cannon Bird) was a complete write-off following a crash-landing near Escalona as a result of engine damage.

He 112 V7 D-IKIK was fitted with a DB 600 engine for its powerplant and was the first of the V-series to have a cut back fuselage. In doing so, it differed visibly from its predecessors.

Left: The V7 in its modified form. External differences included: antenna mast, altered radiator and wings.

Since the V7 served as an engine testbed, the external appearance of the machine changed accordingly. If the aircraft were actually fitted with the two MG 17 guns implied in this photo, these would not have been compatible with the three-blade propeller. It is possible that in this case either no gun or another type of weapon has been fitted. Among the people shown in the picture are Dr. Hertel (third from left) and Ernst Heinkel (seventh from left).

V8, Werknr. 1954, registration D-IRXO (listed as D-IRNO in the type inventory listing of 21 June 1937), with DB 600 A and three-blade Schwarz propeller. Same fuselage style as A-series. Used as testbed for various engine and cooling system designs. Armament consisted of 20 mm MG C 30/L cannons, which eventually were deemed to be "unsuitable for production" (apparently too heavy).

V9, Werknr. 1944, registration D-IGSI, RLM contract from 15 October 1936 as replacement for the destroyed V2. First prototype for B-series. Powerplant was Jumo 210 E, Junkers Hamilton constant speed propeller. In the meantime, as the Luftwaffe's official standard fighter, the Me 109 was fitted with the 1,175 hp DB 601 A. The RLM's Technisches Amt did not approve this engine type for the He 112, since production barely met the needs for the Me 109, Me 110 and He 111. In the Summer of 1937 D-IGSI went to

Switzerland, where it was demonstrated before experts of the Fliegertruppe. Hauptmann Schalk flew the V9 in evaluation flights for the Austrian Bundesheer in November of 1937. The aircraft was sent to Spain in April 1937, coded 8•2 and flown by Harro Harder. Supplied to Hungary in early 1939, where it was lost on a test flight on February 14th, 1939.

V10, Werknr. 2253, registration D-IQMA, first prototype for the E-series (export model of the B-0). Powered by DB 600, later DB 601. First flight in 1938. Heinrich Beauvais flew this machine on 1 September 1938 in Rechlin. Aircraft lost on a test flight.

V11, Werknr. 2254, registration D-IRXS, prototype for B-0 series with two 20 mm Oerlikon/Mauser M cannons (MG 151s) in the wings. First fitted with DB 600, then re-engined with DB 601. Planned as export variant and delivered to Japan, where it was designated as the A 7 He 1.

Above and below: Also fitted with a DB 600 engine, but with oval fuselage: He 112 V8 D-IRXO. The MG FF manufactured by Oerlikon was the 20 mm cannon installed in the engine, since the MG C/30 L had in the meantime been declared unsuitable for fitting.

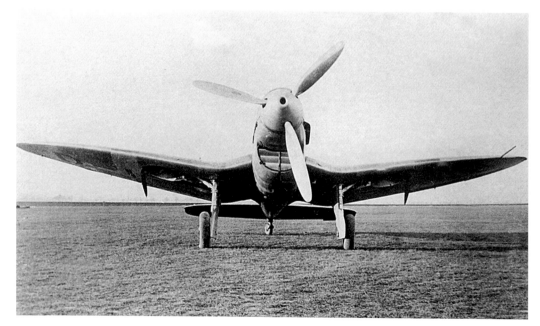

He 112 V9 D-IGSI served many masters: it was tested in Austria, demonstrated in Switzerland, flew as 8•2 in Spain and was eventually delivered to Hungary.

Right: V9 with engine access cover removed during a test runup.

The same aircraft is shown here in flight.

D-IGSI in the summer of 1937 in Switzerland, which was involved in negotiations for the export of the He 112. Access covers for the guns have been opened and a pilot from the Swiss Fliegertruppe is being given an explanation of the aircraft's technical details.

Left: V9 in the Spanish Civil War: marked with the code 8•2, the swastika does not conform to the regulations. It was the "personal marking" of Hauptmann Harro Harder.

Another photo of V9. Notice that, for the purpose of synchronizing the MG 17s, the airplane has been refitted with a two-blade propeller. It also lacks the antenna mast.

The He 112 V10 lost something of its elegance after being fitted with the DB 601 engine, but the aircraft's performance was now on par with the Me 109.

Emergency landing of He 112 V10 D-IQMA during a test flight.

Above and below: A DB 601 engine was also fitted to the He 112 V11 D-IRXS, making the aircraft closely resemble the V10. It was said to have been delivered to Japan, where it was flown under the designation A 7 He 1. The V11 is the last known prototype.

No further V-series prototypes are known to have existed. The following examples are from the A-0 series, which initially was to have comprised twelve aircraft. But, after the A-04, the next four were "skimmed off" for prototype construction (V5 through V8) and continued production of A-series machines ceased afterwards.

A-01, Werknr. 1955, registration D-ISJY, preliminary work begun on 13 January 1936.

A-02, Werknr. 1956, registration D-IXHU, preliminary work begun on 13 January 1936. Flown by Heinz Beauvais on 25 June 1937 in Rechlin.

A-03, Werknr. 1957, registration D-IZMY, powered by Jumo 210 with Schwarz propeller. This machine was exhibited at the IV International Flying Meet in Zürich-Dübendorf from 23 July to 1 August 1937. Due to insurmountable problems with the undercar-

riage it did not participate in any of the competitions, instead being flown by Gerhard Nitschke exclusively in a demonstrative capacity. In October of 1937 the A-03 was part of the German exhibit at the Milan Air Exhibition.

A-04, Werknr. 1958, registration D-IXEU, late 1936/early 1937 was fitted with a Mauser cannon (MG 151, 15 or 20 mm caliber) firing through the propeller hub.

Around 80 aircraft of the B-series were built, all having the "cut fuselage" and enclosed cockpit. In comparison with the B-0, the B-1 had an improved exhaust system. Both variants were primarily designed for the export market (also designated E) with different powerplants and armament, depending on the needs of the buyer and/or authorization by the RLM. It is possible that other versions occasionally mentioned in publications come from this series. The B-machines were fitted with variable pitch VDM propellers.

D-ISJY, the first aircraft from the small A-0 production run. Construction began in January 1936. The archivist has erred here, writing "He 112 V3" on the photo.

Right: A-01 in flight. The retracted landing gear is fully covered. The design bears a passing resemblance to the British Spitfire single seat fighter.

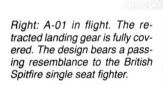

D-ISJY in its original version. In the Third Reich, in-flight pictures of this aircraft were retouched to show the aircraft with a cut back fuselage, giving the impression that the prototype was modified.

He 112 A-03 D-IZMY during the IV International Flying Meet in July of 1937 in Dübendorf, Switzerland.

The same machine as part of the German display at the II International Air Exhibition in Milan in October of 1937.

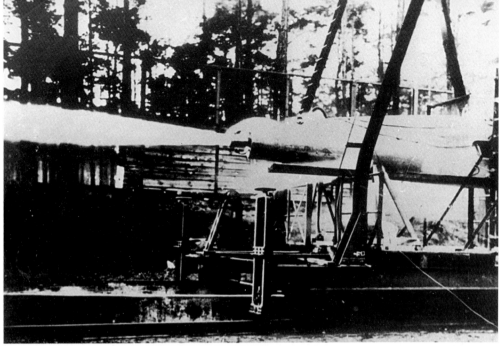

Left: A rocket engine like the one installed in the He 112's fuselage, seen here on the test stand.

Left below: Here the V4 serves as the basis for the alteration. This is not an original picture of a real He 112 rocket powered airplane.

Below: An He 112 with additional rocket motor in flight (propeller running).

The style and location of the Balkan crosses lead one to believe that these were later trials (around 1939).

He 112 B-0s destined for export, in a lineup at Rostock Marienehe.

TYPICAL DAY FOR A TEST PILOT...

Heinkel test pilot Friedrich Ritz flew all kinds of aircraft for this company. He also carried out several flights in the He 112—not always without incident, as the following narrative illustrates. This is taken from his autobiography "Mein Leben", which unfortunately was never published.

Friedrich Ritz recalls: The work at Heinkel had many facets and was interesting beyond imagination, although not entirely without its dangers. Once I was flying an He 112 (note: the V2 is meant), a single seat aircraft conceived and built as a competitor to Messerschmitt's Me 109, in an extended evaluation program and when I went to end my evaluation checkout at low level the engine suddenly cut out. I had no time to look for a suitable emergency landing site, but I rapidly extended the undercarriage (only later did the realization occur that a belly landing was safer in such cases!) and landed in a beautiful paddock. Unfortunately it was a bit too short and the aircraft immediately nosed over in an adjoining plowed field.

The cockpits at the time were still open, and my head was pushed down into the soft earth, forcing me to use my hands to scrape out a hole in order to get fresh air in order to breathe. After about a half an hour had passed, some farmers who had hurried out to the crash site dug me out of my precarious situation.

The heavy pressure on my head cracked two of my neck vertebrae, so I had to give up flying for a few weeks and take it easy. And what was the reason for the engine cutting out, the blame for which they would have liked to have pinned on me (selecting the wrong fuel tank!)? The selector lever for the wing tanks had a switch attached to it which operated the switch linkage: 0-1-2-1+2, equating to 0 = switched off, 1 = left tank, 2 = right tank, 1+2 = both tanks (normal setting). This switch had two arms of unequal length. Due to a construction oversight this switch was installed backwards, so that at the lever's 1+2 position only the right tank was in fact empty; when the aircraft (which had no further damage) was recovered, it was shown that the left tank was full. It was because of this finding by the investigation team that I was, of course, rehabilitated. The previous flight had been flown by Udet, who was fortunate that he didn't fly longer than the fuel supply in the right tank permitted.

After this accident, which turned out not to be that serious, the switch lever was subsequently modified so that it could only be installed in the correct manner.

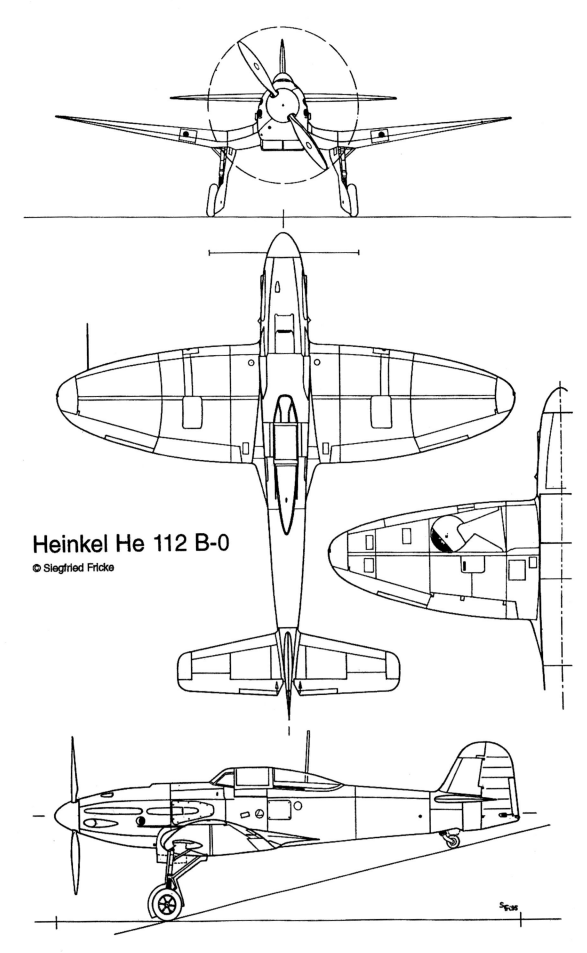

Heinkel He 112 B-0

© Siegfried Fricke

Multi-view drawing of the He 112 B-0 in 1:72nd scale

Left: An He 112 B-series shortly before its completion. It is only lacking the appropriate registration and national markings.

Right: A B-series machine in flight with its radiator extended (upper picture) and retracted. The canopy rear section shows bracing just below the antenna mast, a feature also seen on some of the prototypes for the B-series.

29

Above: A lineup of He 112 B-0s for export. This type was confiscated by the RLM during the so-called Sudeten crisis and absorbed into the regular Luftwaffe.

Below that: D-IRNH just before lifting off.

Left: He 112 B-0 D-IEFC during the International Air Display in Helsingfors from 14-21 May 1938.

Right: Vibration testing on an He 112 B-0.

Below: This He 112 B-0 D-IQRC was demonstrated for the Dutch military aviation in Soesterberg for eight days. However, despite using a more powerful engine (Jumo 211) Heinkel was unsuccessful in motivating the Dutch to purchase this type.

THE He 112 IN SPAIN

As has already been mentioned in the overview of prototypes, Heinkel's He 112 was also flown in Spain. The following abbreviated overview is based in large part on the research of the experts Juan Arráez Cerdá and Harold Thiele:

At the end of November in 1936 the V6 arrived by ship in Cadiz. It was assembled and flown at the airfield in Tablada near Seville. Live firings were conducted here using the 20 mm GB C/30 L cannon firing through the propeller hub. The aircraft subsequently was coded 5•1 and assigned to Versuchs-Jagdstaffel VJ/88 of the Condor Legion. The "5" preceding the "•" was reserved for He 112 aircraft. Günther Radusch began operational testing here on 6 December 1936. It was quickly discovered that attacking armored targets without specialized armor-piercing ammunition was virtually ineffective.

After several sorties flown against ground targets (tanks, trains, vehicles) the machine was lost on 19 July 1937 when Uffz. Schulz made a crash landing due to engine damage shortly before reaching the airfield at Escalona. The wreckage was transported back to Germany to be examined.

Hauptmann Harro Harder brought He 112 V9 to Spain in early April 1938. It was given the code 8•2 and flew operations with fragmentation bombs and three MG 17s. The "8" in front of the "•" actually was assigned to the Russian Polikarpov I-15 Chato biplane, and it is not known why the V9 was not coded with the "5•".

The German government offered the Spanish the He 112 B-0 production variant. After the famous Spanish fighter pilot Garcia Morato put the V9 through its paces and found nothing wrong with the machine, Germany delivered 18 aircraft of this type (two at the end of November 1938, six in January 1939 and ten in April that year). These were given the codes 5•51 through 5•68 and were assigned to the 2nd squadron of the fighter group 5-G-5. Having flown mostly biplanes for quite some time, the Spanish pilots were very impressed with these new aircraft. Compared with the aircraft they'd been flying, the He 112 was for them "a whole new world"...

On 17 January 1939 the first He 112 B-0s were cleared for flight and on January 18th they flew to their airfield at Balaguer. Combat operations began on the 19th of January and Garcia Pardo flushed out an enemy I-15 while on patrol, which he promptly shot down—the first and only kill by an He 112 in the Spanish Civil War. Instructions were later given to protect the ground troops and avoid air combat unless attacked first.

With the Spanish Civil War drawing to a close, the He 112s were deployed to Matacán, then to Grinon and finally to Almaluez. On 28 March 1939 a *Kette* (three-ship formation) took off to observe the Nationalists' march into Madrid from the air. Full of enthusiasm, Rogelio Garcia made a barrel roll at low altitude over the parking area, touched the ground while inverted and crashed. The pilot was killed and the aircraft completely destroyed. At the same time Garcia Pardo staged a mock dogfight with one of his squadron mates, went into a spin and crashed.

Putting a He 112 B-0 together in Spain. On the back wall of the hangar the words "Prohibido fumar" give indication of the existing ban on smoking.

Here, too, the pilot was killed instantly and the aircraft 100% destroyed. Two of the best Spanish pilots had lost their lives at almost the same time, practically at the end of the civil war.

On 1 April 1939 the remaining He 112s were transferred back to Grinon, and on 23 May to Léon. There a reorganization was carried out and the He 112s were transferred to Sania Ramel near Tétouan (then in Spanish Morocco), followed by a deployment to Nador/Tauima near Melilla as 1 Sq. of Grupo 27 of Regimento Mixto No. 2.

On 15 July 1939 Jorge Luis Muntados suffered an engine failure while flying over the Strait of Gilbralter and attempted an emergency landing on the beach near Estepona with his gear extended. The machine flipped over, and the pilot drowned in water approximately a meter deep because the canopy couldn't be opened. The aircraft itself was only slightly damaged.

At the beginning of 1940 Spain occupied the "internationalized" city of Tanger in order to "protect its neutrality". For security purposes a *Rotte* (two-ship formation) of He 112s were stationed at the airfield in Tanger. Up until 1942 no significant activity was noted. Then the Allies landed in North Africa and constantly began violating Spanish airspace without the Spaniards being able to do anything to stop them. For example, on 8 November 1942 the He 112s couldn't prevent it when paratroopers were accidentally dropped near Melilla.

On 3 March 1943 the Spanish pilot Miguel Entrena Klett, while flying He 112 5•65, succeeded in hitting a P-38 Lightning with his two MG FF cannons (it was discovered later that the two MG 17s did not even have ammunition). Trailing smoke, the pilot of the Lightning jettisoned his drop tanks and made a belly landing on the banks of the Muluya River, the border between Algeria and Morocco. By the next day the Americans had recovered the airplane. The Spaniards "captured" the Lightning's drop tanks, which showed several 20 mm holes. The order then went out not to pester Allied aircraft any more.

Above and below: He 112s just after arriving in Spain and being assembled. The camouflage was probably applied back at the Heinkel works. The aircraft carry no additional markings.

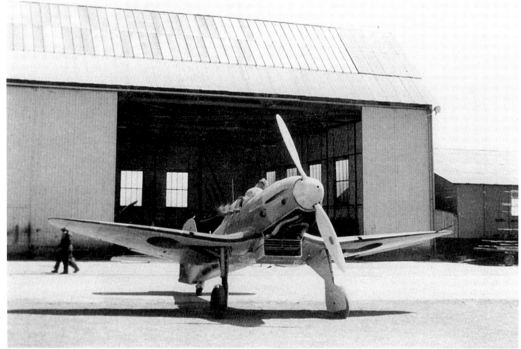

Above and below: The first He 112 B-0s have arrived in Spain, seen here after being assembled in León. The man in the Tirolean hat below is a Heinkel factory pilot.

Miguel Entrena Klett flying over Spanish Morocco. Flying this plane, he forced down a Lockheed P-38 Lightning.

Left: An He 112 B-0 of 1Sq. (Grupo 27) at the Haya airfield near Jerez de la Frontera.

An He 112 B-0 of Grupo 27 in Melilla, Spanish Morocco, in February 1940.

Forced landings: the area surrounding Nador/Tauima was the scene of frequent crashes, such as the one depicted above, which took place on 17 May 1940, or left, in August 1942. Nothing is known about the bottom photo.

The He 112 in Spanish service. The aircraft entered service wearing an overall pale gray finish, which was later superseded by a splinter scheme in brown, green, and gray. Top photo: The emblem of the Spanish fighter squadron 5-G-5 appears on the fin of this He-112 B-0. It was originally the personal emblem of the Spanish fighter ace Garcia Morato. Literally translated, the motto "Vista, suerte y al Toro" means "Search for luck and the bull" or "With a keen eye and some luck—at the bull." The tip of the vertical fin was painted yellow up to aircraft 59 and red from number 60 on, for reasons unknown.

The war was at an end. Franco's dictatorially governed Spain was economically on its last leg. Due to spare parts shortages, fewer and fewer He 112s took to the skies. On 8 April 1946 a crash landing was made near La Restinga, resulting in the death of the pilot José Luis Alvarez and the total loss of the airplane. Emilio Herrera logged his last flight in an He 112 on 23 September 1947. The few remaining aircraft were sent for a short time to the fighter pilots' school in Morón. On 15 July 1952 there was another accident with an He 112, which cost the life of pilot Vinicio Gutierrez Gil, who was attempting to ferry what was probably the last He 112 to the maintenance facility at Logrono.

BRIEF SPELL WITH THE LUFTWAFFE

During the so-called Sudeten crisis of 1938 the RLM confiscated all He 112 B-0s planned for export. These served for a short period with IV/JG 132. Here follows a brief prelude:

On Friday, 1 March 1935 the Luftwaffe was unmasked and now officially became the third branch of service alongside the Heer (Army) and Marine (Navy). On 14 March 1935 the "Reklamestaffel Mitteldeutschland" (Advertising Squadron Central Germany) now became Jagdgeschwader 132 "Richthofen", outfitted with Arado Ar 65s and Heinkel He 51 aircraft. In February of 1937 II/132 was the first Jagdgruppe formed with 25 Messerschmitt Bf 109s.

From 1 July 1937 the IV/132 was stationed in Werneuchen, later the Gruppe operated from Oschatz. After 29 September 1938 (Case "Green", the occupation of the Sudetenland by the German Wehrmacht) JG 132 was equipped for operations as follows: HQ, I and II Gruppe with Bf 109 Bs, III Gruppe with Ar 68s/Bf 109s and IV Gruppe with He 112 B-0s. This Gruppe deployed to Karlsbad on 6 October 1938 and to Trübau in Moravia on 3 November 1938, where it was redesignated I/JG 331. Finally, on 3 February 1939 it was transferred to Breslau-Schöngarten and renamed I/JG 77. By this date at the very latest the last of He 112s would had to have been returned to Heinkel. Staffelkapitän Wolfgang Falck expressed his disappointment in having to give up the He 112s again. He and his pilots preferred flying the He 112 vice the Me 109, which was not considered as robust a design.

A lineup of He 112 B-0s in Luftwaffe markings during the Sudeten crisis.

In Luftwaffe colors and jacked up for aligning the guns. This view shows how elegant the aircraft appeared in its B-version.

He 112 B-0, operating in Luftwaffe service from Werneuchen in 1937/38. Heinkel factory markings of HE+JA have been painted on the wing undersides.

An He 112 B-0 of IV/JG 132. The Jagdgeschwader found itself in Werneuchen in 1937/38, and from September 1938 was stationed at the airbase in Oschatz.

He 112 B-0s in flight during their period of service in Werneuchen (1937/38). Notice that the Balkan crosses have been applied on the extreme outer edges of the wings, corresponding to regulations in effect before 1939.

A row of He 112 B-1s. This variant was exported to Romania.

Last minute preparations before being ferried abroad.

EXPORT WITH RESTRICTIONS

The ignorance of those in power during the Third Reich mentioned in the introduction not only had its effect on the various design proposals, but also on foreign export contracts as well. On the one hand there was considerable dependence on foreign currency, but on the other hand truly new designs were seldom approved for export.

An "exposé for discussing export matters" prepared by the Heinkel firm on 18 August 1938 explores this paradox. It states that the Reich's Minister of Aviation, General Göring, in his speeches before the leading figures of the aviation industry repeatedly stressed the importance of cultivating exports for their products.

It was explained that at the time the aviation industry was virtually taxed to the limit by Reich contracts and the few export contracts would mostly have to be filled without making a profit—despite that, one is aware of one's duty. According to the argument, if export opportunities were to be eliminated then this area would be lost to foreign industry—in a period of time in which many countries were laying the foundations for their own air forces. Not to mention the useful ties which could be established with these countries as a result...

In the years 1937/38 many possibilities for export were undermined by restrictions "from above". The exposé states: "During the course of the last few months we have been forced to watch how Italy, which lives under a similar political situation as Germany, by being able to accommodate all foreign requests at an alarming rate—despite products which are no match for ours—has made inroads into countries which formerly belonged to our customer base." End of quote.

Further examples include: through license sales a large batch of He 112s was to have gone to Yugoslavia. It should be noted here that the German general consul Franz Neuhausen, "fully authorized special emissary of the minister president Generaloberst Göring for Yugoslavia in conjunction with the four-year plan", relayed to the aviation industry economic group on 16 April 1937 that Yugoslavia was interested in purchasing 30 He 112s. 12 were to be pulled from the assembly lines and the remaining 18 were to be acquired as part of a credit package. The deal would have gone through secured if a speed of 500 km/h could have been guaranteed for the aircraft. This would have been entirely possible if the Jumo 210 G fuel injected engine had been fitted in place of the Jumo 210 with carburetor. However, this engine was never approved. Then the engine was cleared for Romania(!), but by that time the British had sewn up the Yugoslavian market.

He 112 B-1 D-IYWE as a postcard picture. Notice the retouched swastika on the tail. At the time, aircraft appearing on postcard were altered, oftentimes quite poorly, in order to disguise their military application or indicate a later time period.

One of the three Hungarian He 112 B-1s. This aircraft, V 303, has yellow painted over the camouflage in the nose, rudder and center fuselage areas.

V 301 of the Magyar Királyi Honvéd Légierö in the spring of 1941 at the Csepel-Budapest airfield, Hungary. P45:

In the first round, the economic loss in foreign currency amounted to 5 million marks.

General Udet personally took an interest in introducing the He 112 to Switzerland. After months of negotiations a preliminary agreement was reached. But the Reich's Ministry of Economy rejected the planned method of payment. Accordingly, the Swiss bought French Morane aircraft. Then the Ministry of Economy said it was prepared to accommodate the Swiss—too late...But there were worse things in store:

Despite buying the French machines, the Swiss gave signals that they were willing to acquire the license for building He 112s and also purchase individual aircraft if immediate delivery were possible. It wasn't possible—all the He 112s destined for export were requisitioned by the RLM!

Example of Hungary: Here, too, negotiations for the sale of the He 112 went on for more than six months. Eventually, the Hungarians were prepared to buy 24 aircraft and obtain the license.

They could have the planes in two to three months. But then the RLM exercised its requisition rights. Ten (!) months after the contract had been issued delivery was promised "in the foreseeable future". The Hungarians considered this a cool rejection and expressed their lack of understanding, particularly since the Italians had offered immediate delivery of Fiat aircraft. At this point Heinkel once again stressed the importance of at least accommodating those countries with which the German Reich had the closest political ties.

In conjunction with other designs and the Asian market, Heinkel attempted to dispel the notion that sales and license approval of the most modern aircraft would jeopardize its own technical advantage. Through ongoing developmental work an advantage of approximately 1.5 years was ensured. On the other hand, there was also the matter of follow-on orders (materials, replacement parts, etc.). All in all, the German Reich had lost considerable credibility on the international scene because of its unreliable export policies.

The RLM reacted to the Heinkel exposé on 27 August 1938. A secret memorandum sent to Heinkel stated that in 1935 a special foreign department of the aviation industry economic group had been set up and infused with a considerable amount of financial means...Due to the high political and military significance, decisions regarding choices, approval or disapproval were to be made on a case by case basis, etc. The specific points in Heinkel's exposé were addressed in a brief and terse manner. Literally:

"1) The military situation does not permit that generous position regarding approvals which you suggest. Military interests take a priority over economic interests. The previous approval policy proved to be necessary militarily—measured in the value of those types found in foreign service and on the front (note: apparently the term "front" was in use as early as August of 1938)

2) It is not possible to completely free exports from politics.

3) The claims regarding Yugoslavia are not applicable. It is a fact that the non-issuance of the contract can be ascribed to other motives than those given by you.

4) Your opinion regarding the failure of the Swiss transaction is not applicable. Fundamental concepts for the Swiss-German business agreement were behind the unwillingness to alter the conditions of payment, concepts which were so important that the Reich Ministry of Economy was forced to reject the Swiss requests even if it meant the loss of the aircraft transaction. In addition, it has been shown that it wasn't economic, but rather political reasons which were the driving force behind contracts being issued to France and England. It is interesting to note that both countries have shown in recent days that they have not upheld their promises, so that Switzerland has of its own volition turned back to Germany for its delivery

5) Even the failure of the Hungarian matter cannot be ascribed to those reasons which you stated, as has been shown in the meantime after detailed inquiries by my ministry. I see from your report that you have not been made aware of the current affairs. By the way, Major Barkacz explained of his own free will in a meeting with the RLM on 22 August 1938 that the rejection of the He 112 was due to technical reasons. Delivery dates had nothing to do with it.

The He 112 in Romanian service. According to data from a Heinkel type sheet published after the war this is a B-1 with minor alterations, designated as a B-2. Unfortunately, the files are missing for this export variant. The photo was taken in the autumn of 1941.

Regarding this matter, I have been informed that you are negotiating with the firm of Manfred Weiss in Budapest for the issuance of a license for the He 112 (without supplying examples) at a price of RM 20,000 together with unit licenses. The Hungarian government has informed me that an official memorandum of understanding is forthcoming regarding the aircraft delivery transactions. I must reject Hungary's common practice whereby the State rejects the type and refuses a license contract and then acquires the license from a firm at an insufficient price, particularly since the hidden interests of this large scale Jewish business are not apparent."

The response went on to discuss other Heinkel designs which are not applicable to this book. Although Heinkel seemed to be a bit dramatic in order to lend emphasis to his exposé, it should be noted that he was quite familiar with the business thanks to many years of export experience and his excellent personal connections. The reasons which the State gave him for the difficulties should be, in part at least, taken with a grain of salt.

Despite all the opposition a number of He 112s were exported abroad. As already mentioned, Spain received a total of 18 aircraft from November 1938 to April 1939, and at least one example was sent to Japan in anticipation of the planned license building and flown as the A 7 He 1. Other sources claim that Japan received over 40 aircraft, but to date there is no evidence (photos, documentation) that this delivery was ever made. By October 1939 Romania had received 30 He 112 B-1s in several batches, which flew missions with the German Luftwaffe during the Second World War. Despite the concerns mentioned earlier Hungary was given licenses for the He 112. That government ordered 36 B-1s on 14 January 1939. Heinkel initially sent the V9 to Budapest-Csepel, where it arrived on 5 February 1939. Nine days later it broke up and was destroyed, but Hungary quickly received a B-1 as a replacement. The large industrial giant Manfred Weiss (building aircraft since 1927) obtained the licenses for building the He 112 and the government placed an initial order for twelve aircraft, of which three (coded V 301, V 302 and V 303) were actually built.

He 112s were almost bought by Austria as well. In November 1937 the decision was made to acquire modern fighters, reconnaissance aircraft and dive bombers. Hauptmann Hans Schalk from the Austrian Fliegerregiment 2 was given the task of inspecting diverse types in Berlin, Bremen, Rostock, Dessau and Köthen beginning on 6 November 1937.

Six Romanian He 112 B-1s can be made out in this photo. The picture was taken in May of 1941 at the Pipera airfield near Bucharest.

Speedy men in the He 112 in conversation: on the left is the Italian record-setting pilot Francesco Agello, in the center is German record holder Hans Dieterle and on the right is Heinkel's chief test pilot Gerhard Nitschke.

The choice in fighters fell upon the He 112. Plans initially called for 42 machines (Service nos. 1001-1042, to be based at Thalerhof). For budgetary reasons the amount was reduced to 36 (total price 5,878,000 RM, spare parts and spare engines not included).

Excerpt from Hauptmann Schalk's report: "I was given the opportunity to inspect and test fly the He 112 and BFW Me 109 fighters. I would like to say in advance that the Me 109 has not yet been cleared for foreign sales. Nevertheless, I flew this type in order to draw a comparison between it and the He 112. Between these two modern aircraft, only the He 112 is worth considering. My judgement is also in favor of this type for the following reasons: 1) In many respects the armament of the He 112 is more advantageous than that of the Me 109. 2) Pressure on all control surfaces is equally strong, while it varies on the Me 109...The reason that the Me 109 and not the He 112 became the Luftwaffe's fighter is because at the time the Luftwaffe was looking for a fighter airplane, and the He 112 was not yet ready for production, and wasn't for another two months. Performance of the He 112 and Me 109 are equally balanced."

However, before deliveries of the He 112 were made, historical events (the annexation of Austria on March 13th, 1938) prevented the transaction from taking place, although the Werkzeugmaschinenfabrik Oerlikon in Zürich did sell 40 pairs of 20 mm Oerlikon FF cannons for the He 112 to the Austrian Air Ministry in Vienna on 9 March 1938.

A total of 11 V-series prototypes, 4 A-0 series and probably 80 export models (serial number block 2001-2080) of the He 112 were built, although not all of the export variants were actually delivered abroad. For example, according to his log book the Rechlin test engineer Walter Baist (engines) conducted a 15 minute factory test flight on 5 June 1940 in an He 112 coded BO+EW. Clues regarding the He 112 aircraft after the beginning of the Second World War are hard to come by; those few examples which were not exported were probably used for various testing purposes.

The He 112's final chapter appears to have been written when the airplane was dropped from Spanish service—in any case, it is unknown whether or not there still exists an example of this type somewhere in the world.

THE LUFTWAFFE PROFILE SERIES

Number 1: Messerschmitt Me 262 by Manfred Griehl
Size: 8 1/2" x 11" b/w and color photographs, color profiles, drawings
52 pages, soft cover
ISBN: 0-88740-820-6 $14.95

Number 2: Messerschmitt Bf 109 G/K by Manfred Griehl
Size: 8 1/2" x 11" b/w and color photographs, color profiles, drawings
52 pages, soft cover
ISBN: 0-88740-818-4 $14.95

Number 3: Heinkel He 219 UHU by Dressel/Griehl
Size: 8 1/2" x 11" b/w and color photographs, color profiles, drawings
52 pages, soft cover
ISBN: 0-88740-819-2 $14.95

Number 4: Focke-Wulf Fw 190
by Manfred Griehl
Size: 8 1/2" x 11" b/w and color photographs, color profiles, drawings
52 pages, soft cover
ISBN: 0-88740-817-6 $14.95

Number 5: Junkers Ju 87A
by Dressel & Griehl
Size: 8 1/2" x 11" b/w photographs, color profiles, drawings
48 pages, soft cover
ISBN: 0-88740-920-2 $14.95

Number 6: Flettner Fl 282
Size: 8 1/2" x 11" b/w photographs, color profiles, drawings
32 pages, soft cover
ISBN: 0-88740-921-0 $12.95

Number 7: Heinkel He 60
by Gerhard Lang
Size: 8 1/2" x 11" b/w photographs, color profiles, drawings
40 pages, soft cover
ISBN: 0-88740-922-9 $12.95

Number 8: Arado Ar 240
by Gerhard Lang
Size: 8 1/2" x 11" b/w photographs, color profiles, drawings
24 pages, soft cover
ISBN: 0-88740-923-7 $9.95